I0475806

Jammu and Kashmir flood crisis: A grim reminder of increasing climate change

Dr. Hemant Pathak

Copyright © 2014 Dr. Hemant Pathak

All rights reserved.

ISBN: 1502530791
ISBN-13: 978-1502530790

DEDICATION

Dedicated to Shri Sainath Maharaj the all omnipotent of world the most

merciful

Also

Tribute to common people of Jammu and Kashmir that fight from

floods bravely.

CONTENTS

Foreword

In the last 10 years, several extreme rainfall events have rocked the country, and this is the latest calamity in that series. Kashmir has witnessed some of the worst episodes in its long history both in terms of human suffering and material destruction. Jammu and Kashmir flood crisis: A grim reminder of increasing climate change; provides a unique insight into the problems our planet faces in terms of clean environment, and what to do about it. This is the only books expressed comprehensive and interdisciplinary focus on, Jammu and Kashmir flood crisis with the multidimensional approach.

This book made of 10 years consistently research on environmental issues, makes it ideal source for students, teachers, industrialist, environmental experts and economists.

This book provides an essential guide to researchers, it offers: various causes of increasing climate change and challenges and experiences in present scenario.

Simply explained, Jammu and Kashmir flood crisis: A grim reminder of increasing climate change is an important book to aware common people how to plan and manage our Environmental resources. This is the real story of the flood tragedy that are hitting Kashmir. We can learn the real lesson behind it can we find ways of dealing with it in the future.

Dr. Hemant Pathak

M.Sc. (Gold medalist), Ph. D.

Assistant Professor of Engineering Chemistry

Indira Gandhi Govt. Engineering College,

Sagar, MP, India

Glossary

Aquatic species Plants and animals found in the water

Aquifer Areas beneath the earth's surface (such as the soil, fractures in the bedrock, or alluvium) where water is stored.

Biomanipulation Change in biological structure by removing and/or stocking living organisms.

Climate prediction A climate prediction or climate forecast is the result of an attempt to produce an estimate of the actual evolution of the climate in the future, for example, at seasonal, interannual, or decadal time scales. Because the future evolution of the climate system may be highly sensitive to initial conditions, such predictions are usually probabilistic in nature.

Climate system The climate system is the highly complex system consisting of five major components: the atmosphere, the hydrosphere, the cryosphere, the lithosphere, and the biosphere, and the interactions among them. The climate system evolves in time under the influence of its own internal dynamics and because of external forcing such as volcanic eruptions, solar variations, and anthropogenic forcing such as the changing composition of the atmosphere and land use change.

Climate change Impact assessment The practice of identifying and evaluating, in monetary and/or nonmonetary terms, the effects of climate change on natural and human systems.

Disaster Severe alterations in the normal functioning of a community or a society due to hazardous physical events interacting with vulnerable social conditions, leading to widespread adverse human, material, economic, or environmental effects that require immediate emergency response to satisfy critical human needs and that may require external support for recovery.

Disaster management
Social processes for designing, implementing, and evaluating strategies, policies, and measures that promote and improve disaster preparedness, response, and recovery practices at different organizational and societal levels.

Early warning system
The set of capacities needed to generate and disseminate timely and meaningful warning information to enable individuals, communities, and organizations threatened by a hazard to prepare to act promptly and appropriately to reduce the possibility of harm or loss.

Ecosystem
An interactive system that includes the organisms of a natural community association together with their abiotic physical, chemical, and geochemical environment.

Flood
The overflowing of the normal confines of a stream or other body of water, or the accumulation of water over areas not normally submerged.
Floods include river (fluvial) floods, flash floods, urban floods, pluvial floods, sewer floods, coastal floods, and glacial lake outburst floods.

Greenhouse effect
The infrared radiative effect of all infrared-absorbing constituents in the atmosphere. Greenhouse gases, clouds, and (to a small extent) aerosols absorb terrestrial radiation emitted by the Earth's surface and elsewhere in the atmosphere. These substances emit infrared radiation in all directions, but, everything else being equal, the net amount emitted to space is normally less than would have been emitted in the absence of these absorbers because of the decline of temperature with altitude in the troposphere and the consequent weakening of emission.

Mitigation
Actions taken to avoid, reduce, or compensate for the effects of environmental damage. Among the broad spectrum of possible actions are

those that restore, enhance, create, or replace damaged ecosystems.

Monitoring   Periodic or continuous surveillance or testing to determine the level of compliance with statutory requirements and/or pollutant levels in various media or in humans, plants, and animals.

Non-point pollution   diffuse pollution mainly from agriculture or dumping grounds. It is difficult to collect for treatment.

Photosynthesis   formation of plant biomass from nutrients with solar radiation as the energy source.

Phytoplankton   free-floating microscopic plants.

Point pollution   polluted water from a defined point. It can be collected as industrial or municipal wastewater and treated by what is often called end-of-pipe technology (environmental technology).

Pollution control   The addition of processes, practices, materials, products or energy to waste streams to reduce the risk posed by pollutants and waste before their release to the environment.

Pollution prevention   The use of processes, practices, materials, products, substances or energy that avoid or minimize the creation of pollutants and waste, and reduce the overall risk to human health or the environment

Reuse   The re-employment of products or materials, in their original form or in new applications, with refurbishing to original or new specifications as required.

1. Introduction

According to ancient Hindu mythology, Environment may be defined as the aggregate of Panch- Tatwa means "the five great elements of nature". The whole universe is made up of these five basic elements namely: Earth, Water, Fire, Air and Space and influences effecting the life and development of an organism.

Development without regard to the ecological balance has led to an environmental crisis in the current scenario. now a day growing evidence of anthropogenic harm in many regions of the earth, dangerous levels of pollution in air, water, earth and living beings. Destruction of its Environment during last decade and a half has no parallel in its entire history.

In September 2014, the Jammu & Kashmir region was hit by heavy floods caused by torrential rainfall. The regions of Jammu and Kashmir in India, as well as POK in Pakistan, were affected by these floods. J&K does not have a flood forecasting system. The Kashmir floods are a grim reminder that climate change is now hitting India harder.

Region grapples with the aftermath of the swirling waters, which have left more than 450 dead and millions homeless.

Many people had died due to the floods. According to the Home Ministry of India, several thousand villages across the state had been hit and 350 villages had been submerged. So far 2,00,000 people have been rescued, including 87,000 from Srinagar city, while leaving over half a million stranded.

Many parts of Srinagar, including civil and army area, were inundated, and vital roads were submerged, by the floods. The current situation in J&K could very well be another manifestation of an extreme weather event, induced by a changing climate.

It is a combination of an intense and unprecedented rainfall event combined with mismanagement (of natural drainage) and unplanned urbanization and lack of preparedness. climate change may have triggered the sudden, intense rainfall that led to the worst floods that the region is facing in over a century.

A projection by PRECIS, a regional climate modeling system, shows that rainfall patterns will change between 2050 and 2100. Rainfall in the sensitive region of the Himalayas will increase by as much as 250 to 500 mm annually while some locations will register an increase of more than 500 mm.

Climate models, analyzed by experts, predict that India will be hit more and more by extreme rainfall events in future and J&K is certainly among the most vulnerable ones due to impact of global warming on Himalayan glaciers. The disastrous damage caused to life and property could have been minimized if the large number of wetlands that once existed in the Valley, had been preserved.

Jammu and Kashmir, has not been prepared to handle such extreme rainfall events. J&K does not have a flood forecasting system. Its disaster management system is also rudimentary.

2.   Back ground

There is a list of extreme events viz. the Mumbai floods of 2005, the Leh cloudburst of 2010. It was barely a year ago that Uttarakhand, another Himalayan state, was devastated by floods. In each of these disasters, thousands have died and the economic losses incurred run into thousands of crores of rupees hitting the country hard.

On September 2, Kashmir was hit by sudden unseasonal and heavy rainfall. It rained more than 200 mm within just 24 hours – four times the average monthly rainfall. This was unexpected and extremely high, and authorities had ignored warnings. The unfolding linkages between climate change and the J&K flood. Scientists say climate models predict that India will be hit more and more by extreme rainfall events. Heavy and very heavy rainfall events in India have increased over the past 50-60 years.

It is proved from continues disasters, climate change is going to affect us more and more worsen in the future. But in echoes of last year's catastrophic flooding in Uttarakhand, critics have accused the administration of being underprepared for the heavy rains.

In 1972, the prime minister of India, Indira Gandhi participated in Stockholm Conference on Human Environment stated as-

The environment cannot be improved in conditions of poverty, unless we are in a position to provide employment and purchasing power for the daily necessities of the tribal people and those who live in around our jungles, we cannot prevent them from combing the forests for food and livelihood; from poaching and from despoiling the vegetation. How can we speak to those who live in villages and slums about keeping the oceans, the rivers and the air clean when their own lives are contaminated at the source.

Major and undesirable disturbances to the ecological balance of the biosphere, destruction and depletion of irreplaceable resources and gross deficiencies harmful to the physical, psychological and social wealth of man in the manmade environment. environmental Pollution is not entirely symbolic of industrial and economic development but can also be indicative of poor technologies, poverty and large population.

In J&K more than 50% of the lakes, ponds and wetlands of Srinagar have been encroached upon for constructing buildings and roads in the past 50 years. The inflow of tons of raw sewage and untreated waste from house-boats and hotels on its banks has converted it into an oversized septic tank. During the reign of the Kashmir Maharajas a well-functioning water drainage system that was synchronous with the natural topography of the region was in place. The same lay neglected in the recent decades.

The banks of the Jhelum river have been taken over in a similar manner, vastly reducing the river's drainage capacity, unplanned urbanization that virtually blocked almost all natural courses of river. Wetlands act as a sponge that retains excess water. Recent floods can be termed as an Ecological Disaster and highlights the urgency to enact a Wetland (Conservation) Act on the pattern of the Forest (Conservation) Act, 1980.

Massive loss of wetlands in the Kashmir Valley over the years. For instance, the famed Dal Lake in Srinagar has seen numerous reclamations all along its periphery in the marshy areas. This has drastically reduced the lake area to just about 1,200 hectares which is almost half of its earlier spread.

The vast expanse of the Wular Lake and associated marshes spread across 20,200 hectares, it now remains restricted to a mere 2,400 hectares. As concrete structures take over wetlands, rivers and streams have lost the ability to carry extra water when it rains heavily. In the last 30 years, nearly 50% of the wetlands in the Kashmir Valley have been encroached upon or severely damaged. British rulers and the former Maharajas of Kashmir used to consider the Wular a buffer for the floods where excess water could be absorbed. The flood channels that used to take the excess water away have been destroyed.

The protection of wetlands will not only help in flood control but will also help in recharging the ground water levels across the country and thus ensure better food security by way of increased water availability.

A 2004 analysis by the Jammu & Kashmir Remote Sensing Centre shows that Srinagar and its suburbs alone have lost 55% of the lakes and wetlands area due to encroachments. Between 1911 and 2004, the area of wetlands went down from 13,426 hectares to 6,407.

The recently-released fifth assessment report of the Intergovernmental Panel on Climate Change projects that India will get more intense rainfall even as the number of rainy days decrease due to changing weather patterns.

The communities residing in the Dal Lake are being held responsible for the lake's pollution, when the truth is that the entire Srinagar's sewage flowed into the Dal till a few years back. Only recently has Srinagar acquired a sewage system that is supposed to handle the city's waste appropriately. The water from the massive catchment comes into the lakes, which are interconnected. More importantly, each lake has its flood discharge channel from where the water spills over for drainage. But over time, we have forgotten the art of drainage, land for building, nothing for water. Dal Lake is virtually at the point of extinction. A massive water body has been reduced to a wide river but without any flow. If the flow had been there, one could have waited to get the original size of the lake back by removing encroachments by various means.

Besides, the siltation of loose soil caused by deforestation and other activities around the catchment areas of the Dal is a major factor polluting the lake, but remains ignored.

Sewage of the Dal dwellers can easily be used up or treated before it enters the lake. Each household and houseboat in the Dal lake can be provided with biogas plants, which will put the sewage to an effective use and provide the people with an eco-friendly and cheaper alternative to LPG.

Our Requires accepting that dealing with climate change impacts is urgent and imperative. Adaptation will require relearning the art and science of water management so that regions such as Kashmir can cope with excess rain in the future.

3. Kashmir floods

The Kashmir floods are a grim reminder that climate change is now hitting India harder. An analysis by the Centre for Science and Environment (CSE) suggests that this could very well be another manifestation of an extreme weather event induced by climate change.

Kashmir floods is one of the worst floods of the century, J&K has been plunged into crisis with people faced with immense difficulties. With the state government facing a heavy flak from all corners of the valley for their late response and mismanagement during the floods, intervention in the fragile environment of the valley may have also escalated the fury after the floods were triggered.

In the last 10 years, several extreme rainfall events have rocked the country, and this is the latest calamity in that series, there has been a severe loss of wetland habitat for various commercial activities.

The Kashmir Valley has one narrow opening towards the west for water from the catchment areas to be drained into the Jhelum. From all other sides, it is encircled by high mountains. The need for a Wetland (Conservation) Act is imperative.

The myriads of lakes and wetlands of the Valley, which acted as sponges, were also well preserved in the past. Considering the importance of wetlands, not just in Jammu and Kashmir, but across India.

Looking back at the cases of previous extreme rainfall events, the scale of disaster in Jammu and Kashmir has been exacerbated by unplanned development especially on the riverbanks. In the last 100 years, more than 50 per cent of the lakes, ponds and wetlands of Srinagar have been encroached upon to construct buildings and roads. The banks of the Jhelum have been taken over in a similar manner, vastly reducing the river's drainage capacity. This is where loss lives and man- made structures have been the most affected.

A study done by B. N. Goswami of the Indian Institute of Tropical Meteorology, Pune, shows that between 1950 and 2000, the incidence of heavy rainfall events (> 100 mm/day) and very heavy events (>150 mm/day) has increased and that of moderate events (5-100 mm/day) has decreased.

The IPCC's 2011 Special Report on Managing the Risks of Extreme Events and Disasters to Advance Climate Change Adaptation presents projections for the period 2071-2100. It points to increasing incidents of more frequent and intense heavy precipitation over most regions.

A wise country should ideally look to the future and take some steps to safeguard it. We will have to see the linkages between climate change and the events such as those unfolding in Jammu and Kashmir. We will have to accept that climate change is going to affect us more and more in the future.

4. Climate change

The earth's climate is dynamic and always changing through a natural cycle. What the world is more worried about is that the changes that are occurring today have been speeded up because of man's activities. These changes are being studied by scientists all over the world who are finding evidence from tree rings, pollen samples, ice cores, and sea sediments.

The U.N.'s panel on climate science said, Climate change is entirely man's fault and limiting its impacts may require reducing greenhouse gas emissions to zero this century.

India is one of the most vulnerable countries. India will also have to proactively work with other countries to reduce emissions to control the warming of the planet. A warmer planet will affect India severely and its poor would be the worst impacted. Causes of the disaster are not tackled through better adaptation to climate change and long-term disaster prevention measures, disasters of this scale may become a regular feature.

The causes of climate change can be divided into two categories - those that are due to natural causes and those that are created by man.

Natural causes

There are a number of natural factors responsible for climate change-

- **Continental drift**

The discovery of fossils of tropical plants (in the form of coal deposits) in Antarctica has led to the conclusion that this frozen land at some time in the past, must have been situated closer to the equator, where the climate was tropical, with swamps and plenty of lush vegetation.

The landmass began gradually drifting apart, millions of years back. This drift also had an impact on the climate because it changed the physical features of the landmass, their position and the position of water bodies. The separation of the landmasses changed the flow of ocean currents and winds, which affected the climate.

This drift of the continents continues even today; the Himalayan range is rising by about 1 mm (millimeter) every year because the Indian land mass is moving towards the Asian land mass, slowly but steadily.

- **Volcanoes**

When a volcano erupts it throws out large volumes of sulphur dioxide (SO_2), water vapour, dust, and ash into the atmosphere. The large volumes of gases and ash can influence climatic patterns for years. Millions of tones of sulphur dioxide gas can reach the upper levels of the atmosphere from a major eruption.

The gases and dust particles partially block the incoming rays of the sun, leading to cooling. Sulphur dioxide combines with water to form tiny droplets of sulphuric acid.

These droplets are so small that many of them can stay aloft for several years. They are efficient reflectors of sunlight, and screen the ground from some of the energy that it would ordinarily receive from the sun. Winds in the upper levels of the atmosphere, called the stratosphere, carry the aerosols rapidly around the globe in either an easterly or westerly direction.

Movement of aerosols north and south is always much slower. This should give you some idea of the ways by which cooling can be brought about for a few years after a major volcanic eruption.

Mount Pinatubo, in the Philippine islands erupted in April 1991 emitting thousands of tones of gases into the atmosphere. Volcanic eruptions of this magnitude can reduce the amount of solar radiation reaching the Earth's surface, lowering temperatures in the lower levels of the atmosphere and changing atmospheric circulation patterns. The extent to which this occurs is an ongoing debate.

- **The earth's tilt**

The earth makes one full orbit around the sun each year. It is tilted at an angle of 23.5° to the perpendicular plane of its orbital path. For one half of the year when it is summer, the northern hemisphere tilts towards the sun. In the other half when it is winter, the earth is tilted away from the sun. If there was no tilt we would not have experienced seasons.

Changes in the tilt of the earth can affect the severity of the seasons - more tilt means warmer summers and colder winters; less tilt means cooler summers and milder winters.

The Earth's orbit is somewhat elliptical, which means that the distance between the earth and the Sun varies over the course of a year. We usually think of the earth's axis as being fixed, after all, it always seems to point toward Polaris. Actually, it is not quite constant: the axis does move, at the rate of a little more than a half-degree each century. So Polaris has not always been, and will not always be, the star pointing to the North. When the pyramids were built, around 2500 BC, the pole was near the star Thuban (Alpha Draconis). This gradual change in the direction of the earth's axis, called precession is responsible for changes in the climate.

- **Ocean currents**

The oceans are a major component of the climate system. They cover about 71% of the Earth and absorb about twice as much of the sun's radiation as the atmosphere or the land surface.

Ocean currents move vast amounts of heat across the planet roughly the same amount as the atmosphere does. But the oceans are surrounded by land masses, so heat transport through the water is through channels.

Winds push horizontally against the sea surface and drive ocean current patterns. Certain parts of the world are influenced by ocean currents more than others. The coast of Peru and other adjoining regions are directly influenced by the Humboldt current that flows along the coastline of Peru. The El Niño event in the Pacific Ocean can affect climatic conditions all over the world.

Another region that is strongly influenced by ocean currents is the North Atlantic. If we compare places at the same latitude in Europe and North America the effect is immediately obvious. Take a closer look at this example - some parts of coastal Norway have an average temperature of -2°C in January and 14°C in July; while places at the same latitude on the Pacific coast of Alaska are far colder: -15°C in January and only 10°C in July. The warm current along the Norewgian coast keeps much of the Greenland-Norwegian Sea free of ice even in winter. The rest of the Arctic Ocean, even though it is much further south, remains frozen.

Ocean currents have been known to change direction or slow down. Much of the heat that escapes from the oceans is in the form of water vapour, the most abundant greenhouse gas on Earth. water vapor also contributes to the formation of clouds, which shade the surface and have a net cooling effect.

Human causes

The Industrial Revolution in the 19th century saw the large-scale use of fossil fuels for industrial activities. These industries created jobs and over the years, people moved from rural areas to the cities. This trend is continuing even today. More and more land that was covered with vegetation has been cleared to make way for houses.

Natural resources are being used extensively for construction, industries, transport, and consumption. Consumerism has increased by leaps and bounds, creating mountains of waste. Also, our population has increased to an incredible extent.

All this has contributed to a rise in greenhouse gases in the atmosphere. Fossil fuels such as oil, coal and natural gas supply most of the energy needed to run vehicles, generate electricity for industries, households, etc.

The energy sector is responsible for about ¾ of the carbon dioxide emissions, 1/5 of the methane emissions and a large quantity of nitrous oxide. It also produces nitrogen oxides (NOx) and carbon monoxide (CO) which are not greenhouse gases but do have an influence on the chemical cycles in the atmosphere that produce or destroy greenhouse gases.

- **Greenhouse gases and their sources**

Carbon dioxide is important greenhouse gas in the atmosphere. Changes in land use pattern, deforestation, land clearing, agriculture, and other activities have all led to a rise in the emission of carbon dioxide.

Methane is another important greenhouse gas in the atmosphere. About ¼ of all methane emissions are said to come from domesticated animals such as dairy cows, goats, pigs, buffaloes, camels, horses, and sheep. These animals produce methane during the cud-chewing process. Methane is also released from rice or paddy fields that are flooded during the sowing and maturing periods.

When soil is covered with water it becomes anaerobic or lacking in oxygen. Under such conditions, methane-producing bacteria and other organisms decompose organic matter in the soil to form methane. Nearly 90% of the paddy-growing area in the world is found in Asia, as rice is the staple food there. China and India, between them, have 80-90% of the world's rice-growing areas.

Methane is also emitted from landfills and other waste dumps. If the waste is put into an incinerator or burnt in the open, carbon dioxide is emitted. Methane is also emitted during the process of oil drilling, coal mining and also from leaking gas pipelines (due to accidents and poor maintenance of sites).

A large amount of nitrous oxide emission has been attributed to fertilizer application. This in turn depends on the type of fertilizer that is used, how and when it is used and the methods of tilling that are followed. Contributions are also made by leguminous plants, such as beans and pulses that add nitrogen to the soil.

- **Carbon emissions**

The climate change brought about by man-made carbon emissions is heating up the atmosphere faster than normal. Consequently a change in global weather patterns and its natural variability, Monsoons are generally confounding natural events that are hard to predict and even harder to pin down. Even then scientists are able to find a change in patterns.

Climate models predict that heavy rain events will increase over the Indian subcontinent. The Inter-governmental Panel on Climate Change reports confirm that climate change will lead to an increase in frequency, intensity, spatial extent, duration, and timing of extreme weather events.

5. Government Efforts

"We want to understand the reasons behind such events and are also exploring international cooperation, as extreme weather events are inter-related with changes in climactic conditions in

North Pole impacting the Indian monsoon," the New Indian Express quoted an official from the ministry of Earth sciences saying.

The government says it plans to expand the country's network of forecasting stations, building nine extra radars in Himalayan cities by 2017.

But it remains unclear whether the national government is willing to amend and enforce planning laws to make sure buildings and communities are better equipped to deal with intense events. Some questions remains about inadequate government regulation of land use, rising levels of deforestation and a proliferation of hydropower plants in the Himalayas.

Most climate models also predict that India will be hit more and more by extreme rainfall events as the world continues to warm in the coming decades. According to the latest analysis by Working Group II of the IPCC Assessment Report (AR5), floods and droughts are likely to increase in India. India will get more rainfall but it won't be spread over a number of days. An increase in extreme precipitation during monsoons is also predicted.

6. Conclusions

India's worked on environment action plan since last 30 years actively for controlling pollution at the national level calls for the implementation of time-bound programmes that entail coordinated interdepartmental strategies.

People of the whole world and the duty of the all Governments to protect environment. Countries such as India to raise the importance of climate change action agreement with the world community.

Articles 48A and 51A of the Constitution of India have cast a solemn duty not only on the State but also on the citizens towards the protection of the environment and conservation of the forests and the wild life.

Causes of devastation following extreme events viz. droughts, floods etc., often complicated, and mismanagement of resources and poor planning also share the blame.

Kashmir's unusually high rainfall was only part of the problem. The state does not have a flood forecasting system or capacity for disaster preparedness.

Most of the natural drainage channels have been destroyed due to utter mismanagement. The destruction has been deliberate and intentional purely for material greed which has surpassed all limits.

The traditional system of flood management was to channelize the water from the Himalayas into lakes and water channels. The Dal and Nageen lakes in Srinagar are not just its beauty spots but also its sponges.

Protection and improvement of human environment is a major issue which effects the well beings of the people and economic development throughout the world.

Increased capacity of forecasting and information dissemination so that people are aware of the dangers and lives are not lost.

It envisages control of pollution from various sources such as industrial, domestic, vehicular, agricultural and noise.

7. References

1. Connecting Biodiversity and Climate Change Mitigation and Adaptation: Report of the Second Ad Hoc Technical Expert Group on Biodiversity and Climate Change. Technical Series No. 41, Secretariat of the Convention on Biological Diversity (CBD), Montreal, QC, Canada, 126 pp.

2. Climate Change 1992: The Supplementary Report to the IPCC Scientific Assessment [Houghton, J.T., B.A. Callander, and S.K. Varney (eds.)]. Cambridge University Press, Cambridge, UK and New York, NY, USA, 116 pp.

3. Climate Change 1995: The Science of Climate Change. Contribution of Working Group I to the Second Assessment Report of the Intergovernmental Panel on Climate Change [Houghton, J.T., L.G. Meira Filho, B.A. Callander, N. Harris, A. Kattenberg, and K. Maskell (eds.)]. Cambridge University Press, Cambridge, UK and New York, NY, USA, 572 pp.

4. Workshop Report of the Intergovernmental Panel on Climate Change Workshop on Impacts of Ocean Acidification on Marine Biology and Ecosystems [Field, C.B., V. Barros, T.F. Stocker, D. Qin, K.J. Mach, G.-K. Plattner, M.D. Mastrandrea, M. Tignor, and K.L. Ebi (eds.)]. IPCC Working Group II Technical Support Unit, Carnegie Institution, Stanford, CA, USA, 164 pp.

5. UNISDR Terminology on Disaster Risk Reduction. United Nations International Strategy for Disaster Reduction (UNISDR), United Nations, Geneva, Switzerland, 30 pp.

6. WCED, 1987: Our Common Future. World Commission on Environment and Development (WCED), Oxford University Press, Oxford, UK, 300 pp.

7. NRC (2010). *Advancing the Science of Climate Change*. National Research Council. The National Academies Press, Washington, DC, USA.

8. Jansen, E., J. Overpeck, K.R. Briffa, J.-C. Duplessy, F. Joos, V. Masson-Delmotte, D. Olago, B. Otto-Bliesner, W.R. Peltier, S. Rahmstorf, R. Ramesh, D. Raynaud, D. Rind, O. Solomina, R. Villalba and D. Zhang (2007). Paleoclimate. In:*Climate Change 2007: The Physical Science Basis. Contribution of Working Group I to the Fourth Assessment Report of the Intergovernmental Panel on Climate Change* [Solomon, S., D. Qin, M. Manning, Z. Chen, M. Marquis, K.B. Averyt, M. Tignor and H.L. Miller (eds.)]. Cambridge University Press, Cambridge, United Kingdom and New York, NY, USA.

9. Solomon, S., D. Qin, M. Manning, R.B. Alley, T. Berntsen, N.L. Bindoff, Z. Chen, A. Chidthaisong, J.M. Gregory, G.C. Hegerl, M. Heimann, B. Hewitson, B.J. Hoskins, F. Joos, J. Jouzel, V. Kattsov, U. Lohmann, T. Matsuno, M. Molina, N. Nicholls, J. Overpeck, G. Raga, V. Ramaswamy, J. Ren, M. Rusticucci, R. Somerville, T.F. Stocker, P. Whetton, R.A. Wood and D. Wratt (2007). Technical Summary. In: *Climate Change 2007: The Physical Science Basis . Contribution of Working Group I to the Fourth Assessment Report of the Intergovernmental Panel on Climate Change* [Solomon, S., D. Qin, M. Manning, Z. Chen, M. Marquis, K.B. Averyt, M. Tignor and H.L. Miller (eds.)]. Cambridge University Press, Cambridge, United Kingdom and New York, NY, USA.

10. Forster, P., V. Ramaswamy, P. Artaxo, T. Berntsen, R. Betts, D.W. Fahey, J. Haywood, J. Lean, D.C. Lowe, G. Myhre, J. Nganga, R. Prinn, G. Raga, M. Schulz and R. Van Dorland (2007). Changes in Atmospheric Constituents and in Radiative Forcing. In: *Climate Change 2007: The Physical Science Basis . Contribution of Working Group I to the Fourth Assessment Report of the Intergovernmental Panel on Climate Change* [Solomon, S., D. Qin, M. Manning, Z. Chen, M. Marquis, K.B. Averyt, M. Tignor and H.L. Miller (eds.)]. Cambridge University Press, Cambridge, United Kingdom and New York, NY, USA.

11. USGCRP (2009). *Global Climate Change Impacts in the United States* . Thomas R. Karl, Jerry M. Melillo, and Thomas C. Peterson (eds.). United States Global Change Research Program. Cambridge University Press, New York, NY, USA.

12. NRC (2002). *Abrupt Climate Change: Inevitable Surprises* . National Research Council. The National Academies Press, Washington, DC, USA.

13. Hegerl, G.C., F. W. Zwiers, P. Braconnot, N.P. Gillett, Y. Luo, J.A. Marengo Orsini, N. Nicholls, J.E. Penner and P.A. Stott (2007). Understanding and Attributing Climate Change. In: *Climate Change 2007: The Physical Science Basis .Contribution of Working Group I to the Fourth Assessment Report of the Intergovernmental Panel on Climate Change*[Solomon, S., D. Qin, M. Manning, Z. Chen, M. Marquis, K.B. Averyt, M. Tignor and H.L. Miller (eds.)]. Cambridge University Press, Cambridge, United Kingdom and New York, NY, USA.

14. UNEP/WMO (2011) *Integrated Assessment of Black Carbon and Tropospheric Ozone: Summary for Decision Makers* . United Nations Environmental Programme and the World Meteorological Society.

ABOUT THE AUTHOR

Dr. Hemant Pathak held positions as Assistant Professor in the department of chemistry, Govt. Indira Gandhi Engineering College, Sagar, MP, India. He had extensive experience in teaching, research and administrative management.

Dr. Pathak received his Ph.D. degree in chemistry from Dr. Hari Singh Gour Central University, Sagar, India and M.Sc. Gold medalist from Jiwaji University, Gwalior. He has published 35 books (including e- books) and more than 50 research papers in reputed International and National journals and received several awards. He is a member of editorial boards and reviewer boards of several international journals and societies. His area of specialization includes Engineering Chemistry, Energy audits and Environmental Pollution management.

www.ingramcontent.com/pod-product-compliance
Lightning Source LLC
Chambersburg PA
CBHW081250170526
45165CB00009B/3269